I0486097

TABLE OF CONTENTS

PREFACE

Very little information has been published with any engineering confidence as to why the pyramids were built. The most popular reason believed by astrologist is that they were built as a structure to bury the body of the Egyptian pharaohs. However, the big problem with this is: There has never been even one Egyptian pharaoh's body found in any of the pyramids.

Also, there were many more pyramids built than there were pharaohs. Other reasons that some people think the pyramids were built are: 1.) sites for sacrifice. 2.) places for worship. 3.) astronomical tools. Also, many astrologists believe the causeway leading up to the Great Pyramid of Giza was used for transporting the body of the pharaoh to inside of the pyramid. To my way of thinking - this is ludicrous .

I have spent several years thinking about why the pyramids were built, and after analyzing the existing internal structure of the great pyramid, I believe, beyond doubt, that I have discovered the precise reason that they were build. This reasoning, with drawings and descriptive write ups are shown in the body of this book.

DEDICATED

This book is dedicated to my three children whom I love deeply

First, my eldest son Richard Charles Phillips, now 56 years old. He lives in Birmingham, Alabama. He is the music director at South Highlands Methodists Church, and teaches music at the Advent School in Birmingham, Alabama. He has taught there for 32 years. He is married to Suzanne Phillips and they have 2 lovely children Lucy, 21 years old, and Gracie, 16 years old.

Second, My only daughter, Tressa Fouts, now 53 years old. She is married to Paul Fouts, and they have one son named Benjamin. He is married and they all reside in the Stone Mountain Area in Atlanta Georgia. They have a successful landscape company in this area.

Third, my youngest son, William Christopher Phillips, now 46 years old. He is a successful HVAC design engineer, and currently resides with his father, and helps his father and step mother, Evelyne Phillips, manage a successful real estate rental company in the Birmingham, Alabama, area.

ABOUT THE AUTHOR

Richard Cary Phillips was born April 5, 1934, in Starkville, Mississippi. At the age of seven, he moved to Birmingham, Alabama, where he attended Glen Iris Elementary School, and later, Ramsay High School, majoring in the technical curriculum, with emphasis on math and science. He played football and four years of baseball.

In 1953, he attended his first year at Auburn University, and majored in Mechanical Engineering. He played baseball while attending Auburn, and in 1957, he received a Bachelor's Degree in Mechanical Engineering.

Volunteering under the Army's Federal Reserve Act, Mr. Phillips spent six months on active duty from September, 1957 to March 1958 at Fort Jackson, South Carolina, and subsequently for the next six years, he served in the US Army Reserves.

In 1958, he married Shirley Dean Golden, from Leeds, Alabama. Their first son was born, Richard Charles Phillips, in September 1959. Mr. Phillips started to work on his first meaningful job with US Pipe and Foundry in Birmingham, Alabama. In 1966, Mr. Phillips resigned from US Pipe and Foundry, and began a career with the aerospace industry.

In 1962, while living in Birmingham, his second child was born, a beautiful girl named Tressa Ann. At the age of twenty-one, she married Paul Fouts in Roswell, Georgia. She has only one son, named Benjamin. They all currently reside near Stone Mountain, Georgia.

Mr. Phillips relocated to Huntsville, Alabama in 1966 to work as a facilities engineer for the Boeing Company on the Apollo Saturn V Program. In this effort, he personally designed all the piping systems necessary for the water-fill piping to the Saturn V rocket's fuel tanks at the DTV test facility. This was necessary in order to evaluate the stresses on the rocket when the fuel tanks were filled with water, which stimulated liquid oxygen and kerosene, the propellant used for the Saturn V rocket. When filled with water, the rocket was vibrated by a huge vibrator at the same intensity as it would occur during an actual liftoff by the rocket. This test was necessary to evaluate if the rocket would maintain its integrity during an actual liftoff to the moon.

Other work with the Boeing Company in Huntsville included designing piping systems at the Bread Board for the Saturn V Rocket. The Bread Board was the test facility that generated computer programs that would sequence the opening and shutting off of physical valves, and piping systems that simulated and actual trip of the Saturn V rocket's flight to the moon and back possible. Mr. Phillips designed and installed much of the piping and valves necessary to make these computer - simulated test runs to the moon and back possible

After one year at Huntsville, Boeing asked him to go to Merritt Island, Florida, where he was a lead facilities engineer on the Apollo Saturn V program. One of the first assignments was to help complete the final facilities work necessary to launch the first Apollo Saturn V rocket in orbit around the Earth. Some of this work consisted of the modification of the cryogenic piping located at launch pad 39A, the launch pad to the moon. Other work included the air-conditioning modification inside the launch pad facility at 39A. Other facilities work included high-pressure hydraulic piping modifications inside the Vehicle Assembly Building, prior to the Saturn V rocket going to the launch pad 39A.

After completion of this work and after the first Saturn V rocket was successfully launched into orbit around Earth, Mr. Phillips was then assigned to work on the Boeing's TIE Program. This program allowed Mr. Phillips to represent NASA and wear the NASA hat while working for Boeing. In this capacity, he worked with the various contractors from the Saturn V program to oversee their interface with other facilities' equipment and piping, with all the various other contractors furnishing their services to get a man on the moon. The TIE program was the abbreviation for Technical Integration and Evaluation, and Mr. Phillips' office was located in the Spacecraft Headquarters Building of NASA. This was the building where the actual spacecraft, called the Lunar Excursion Module (LEM), was located that was to land a man on the moon. Also, this building contained the actual Service Module that was to orbit the moon and bring the astronauts back to earth. Much of Mr. Phillips' work was fieldwork necessary to determine the various contractor facilities interfaces connected to the Lunar Excursion Module and the Service Module.

Other work on the TIE program, performed by Mr. Phillips, consisted of fieldwork of checking contractor-piping interfaces on the launch pad 39A to see that all interfaces were identified before the launch of the Saturn V rocket, Lunar Excursion Module, and Service Module.

After his field investigative work, he would generate description write-ups and schematic drawings necessary for NASA personnel to troubleshoot exactly where a problem existed and which company was responsible for the problem. Should any problems arise with any of the facilities equipment necessary to get the Saturn V rocket to the moon, Mr. Phillips furnished NASA with the information as to which contractor was responsible for the problem, and with whom that contractor interfaced.
For Mr. Phillips' contribution to aid NASA information in the description of all contractor interfaces and schematic drawings, to land a man on the moon, Mr. Phillips was awarded to the Apollo Saturn V Roll of Honor.

The following is a written evaluation of Mr. Phillips' performance, given by Mr. Phillips' manager of the Apollo Saturn V rocket of the Boeing Company. Also this written information about Mr. Phillips, in his effort in getting man on the moon, has been documented and recorded in a book titled *Apollo/Saturn V Roll of Honor,* located in the Library of Congress, Washington, DC and the Smithsonian Institution, Washington, DC. The description of his work as stated by Mr. Paul Bromstad, Boeing Facilities Manager, is as follows:

"Mr. Phillips researched and analyzed complicated mechanical Ground Support Systems and reduced them to accurate, readily understandable schematic forms. These schematic layouts were published in the SCO Launch Complex-34 and Launch Complex-39 Operational and Maintenance Interface documents, providing NASA/SCO management visibility for timely resolution of problems related to these systems. In the foregoing assignment, he displayed initiative and resourcefulness in researching , field checking, and coordinating system details with the proper source in order to develop concise, accurate, and reliable engineering information. The results of his work have been outstanding in terms of customer acceptance and satisfaction."

Mr. Phillips first son Richard Charles Phillips, now fifty-six years old, is a professor of music at a private school in Birmingham, AL. He won a full scholarship at Birmingham Southern College, and was recognized in *Who's Who in America*. He has earned a master's degree in music. He now resides in Vestavia Hills, AL with his wife Suzanne and two daughters Lucy and Gracie.

In 1969, Mr. Phillips's second son, William Christopher Phillips, was born in Merritt Island, Florida. Chris is now forty-six years old. He married Leigh Ann in 1994, but they are now divorced. He has only one daughter, Abigail who is now twenty years old. Chris worked most of his career in Birmingham, AL as a design engineer in the mechanical field. All of his work is computer generated.

Mr. Phillips has noted that he is very blessed to have wonderful children and grandchildren, and he is extremely proud of all of them.

After completion of his work with the Boeing Company in helping to get a man on the moon, Mr. Phillips returned with his wife and three children to Birmingham, AL, where he obtained a position with Southern Services, Inc., designing piping systems for electric power plants for the Southern Company.

After a short time doing design piping for Southern Services, Mr. Phillips was promoted to be in charge of the heating, ventilation, and air conditioning (HVAC) of all the nuclear power plants that were being designed for the Southern Company. His first assignment was the supervision of the design and construction of the HVAC equipment for the E.I.

Hatch Nuclear Power Plant in Georgia. This was the first nuclear power plant constructed in the state of Georgia.

After the successful completion of this work, he supervised the engineering and design of the HVAC equipment, ductwork, and specifications of the Vogel Nuclear Power Plant in Alabama. Later, this name was changed to the Central Alabama Nuclear Power Plant.

After the completion of this work in 1973, Mr. Phillips was offered a large monetary increase to go to another nuclear power engineering company, EBASCO, located in Norcross, GA to do HVAC nuclear engineering. In 1973, Mr. Phillips moved with his family, to Roswell, GA to work for this company. During the next couple of years, Mr. Phillips performed HVAC engineering on the Sharon Harris Nuclear Power Plant. He was the project engineer on the HVAC for the design of the radwaste building and supervised the design of all the ductwork, and he wrote the specifications on all the equipment.

After two years with EBASCO, he resigned the company to pursue personal engineering projects and research regarding this project.

Mr. Phillips and his family survived during this period of research by having some income from real estate rental property he owned and by the help of his wife managing a restaurant that Mr. Phillips leased in Roswell, GA. In essence, Mr. Phillips took a sabbatical from his career in nuclear engineering and devoted the majority of his time in research.

After his short sabbatical with fulltime companies, he elected to start a new engineering career by contracting out his services to various companies, as a HVAC consulting engineer. Some of the companies where he worked as a consulting engineer are as follows:

1. 1980-1981   Herry and Herry, an architectural and engineering firm Atlanta, GA.

2. 1981-1982   Robert & Company, an architectural and engineering firm in Atlanta, GA.

3. 1982-1983   Frito-Lay, a food processing company in Dallas, TX.

4. 1983-1984   Kimberly Clark, a paper manufacturing company in Roswell, GA.

5. 1984-1986   Cadre Corporation, an engineering firm in Atlanta, GA.

6.	1986-1988	TVA, an electrical utility company in Knoxville, TN.
7.	1989-1989	Rust Engineering, an engineering and design company in Birmingham, AL
8.	1989-1991	The Boeing Company, an aerospace company in Huntsville, AL.
9.	1991-1991	Herndon Engineering, a mechanical design company in Birmingham, AL
10.	1992-1992	Adtechs Corporation, a facilities design company in Herndon, VA.
11.	1992-1993	Bechtel National, an engineering firm in Titusville, FL.
12.	1993-1993	Boyle Engineering, Inc., an engineering firm in Orlando, FL.
13.	1993-1999	McCauley & Associates, an architectural firm in Birmingham, AL.
14.	1999-Present	Bought and leased out homes in the Birmingham, AL.

In 1983, Mr. Phillips met Evelyne Marlowe, while working for Kimberly Clark. She was forty-one years old and French born. She had married an American soldier while in France and moved to the US in May of 1964. She became a US Citizen in 1994. Evelyne's job was in retail sales (diamonds and fine jewelry) when they met. She had three children by her first marriage. Her youngest daughter lived with her at the time, named Pamela, who was thirteen years old. She graduated from the University of Alabama (suma cum laude) and shortly after, earned two masters' degrees. She is currently a director for physicians' services for HCA in Nashville, TN. Evelyne's oldest daughter, Cynthia, is married to an industrial engineer, has two grown daughters (Jessica and Megan) and she lives in Tullahoma, TN. Evelyne's only son, Gregory, is a fire fighter in Largo, FL, and is a war veteran. He was married and has one son named Benjamin.

On May 3, 1984, Mr. Phillips and Evelyne Marlowe were married by the Reverend Paul Walker of Mount Paran Church of God in Atlanta, GA. They have lived happily together for thirty one years.

In September 1986, Mr. Phillips and Evelyne moved from Roswell, GA to Birmingham, AL into a home he previously owned.

In 1994, Mr. Phillips and his wife began purchasing investment homes in the Birmingham, AL area, and renting them out to provide an income for now, and later for their retirement. During the period from 1994 to 2004, Mr. Phillips and his wife purchased and rented out twenty homes.

In the writing of these published works, Mr. Phillips would like to give special thanks to his dear friend Karen Vowells, who never lost faith in him. She edited and critiqued much of the work that he did, and supported him in every way she could to get this "much needed" information out to the world. He said that he had been blessed to have her valuable friendship for over the past 63 years.

Other books and DVDs published by Mr. Richard Cary Phillips are as follows:

"The Answer to the Propulsion of Flying Saucers" . This book describes exactly how the so-called flying saucers are propelled. It shows calculations to prove how tremendous forces are generated to go to Mars in 9 days instead of 9 months. It has patent US 7,634,903 attached to the back of the book. Book contains 101 pages.

"How a UFO Could Capture a Boeing 777 by the use of Quick Silver". This book describes how a UFO could attach to a Boeing 777 and carry it up into outer space, or even to the moon or Mars with no problem. The manner in which it can be captured is in the specifications of the book, and is verified by patent US 7,634,903 contained in this book. The book contains 133 pages.

"How a "Free Energy" 400 Horsepower Automobile Engine can run Indefinitely" This book shows general arrangement, and detailed drawings, how it is possible to design an engine to produce a "Free Energy" a 400 horsepower automobile engine that will run indefinitely. For its power, it uses the principle of my US Patent No. 7,634,903 for the propulsion of this engine. This book is only 60 pages, but is very technically intense, and would be comprehended only by special people with above average IQ,s. It is not for people, who are mentally challenged with a new innovative technology to purchase.

"How UFO"s fly - Fully Explained " This a two hour DVD that is available for those people who find reading difficult.   I explain, with a narrative, and model props, how UFOs are propelled. I show explicit passages in the Bible (Kings James version) where Ezekiel describes in over 10 passages, that are directly related to the physical design of the so-called "Flying Saucer"

The DVD explains the three distinct methods of flight in which the UFO can utilize, 1.) It can hover in our atmosphere for hours, utilizing the spent propellant from the craft. 2.) It can be propelled in outer space to fly at 10's of thousands of miles per hour. 3.) It can maneuver in our atmosphere, and outer space, in the same manner as our helicopters.

CHAPTER 1 - SPECIFICATIONS OF THIS DISCOVERY

Now that I have a patent no. 7,634,903 B2, I know that it is the same propulsion system that is used by the so - called "Flying Saucers" that are seen today. I also believe that UFOs have been around for thousands of years, and it is very obvious to me who built the pyramids. This engine that I have a patent on can lift hundreds of tons upward to build such pyramids that exist today.

As a matter of fact, as noted, I am a registered professional engineer, and I do not think the pyramids could be built today with our modern technology. The 60 ton blocks would have to be lowered from the sky with something like a helicopter, but our helicopters today do not have the capacity to lift such weights. Also, I do not believe the Egyptians had cranes with 450 foot booms, to lift those weights at the boom angle of about 40 degrees to put the last blocks at the top of the pyramid. Granted, cranes could be built to do this, at a cost of millions of dollars, but they are not available today and certainly were not available when the pyramids were built.

According to my patented invention, it is my belief that the beings who flew the so-called UFOs built the pyramids, and not the pharaohs. It's quite obvious that they were not built for the pharaohs' burial tombs, since no mummified pharaoh has ever been found in any pyramids whatsoever.

Now, to get to the real subject matter of this book is as follows: If you lived in Egypt with nothing but a hot desert all around you, what would be the one thing that would bring life to the area, and make it livable? What would be worth its weight in gold if you did not have it? The answer to these questions, to a man dying of thirst in the desert, would be water. Yes, water! This would be like gold.

This will not be a long book on this subject, since it will be very simple once you understand the intelligence of these beings who built the UFOs. They also have the capability to build a tremendous pyramid with a purpose behind it. That purpose is to pump water to the desert for drinking, washing and mainly irrigation. You see, our historians say that Egypt flourished with tremendous agriculture because of the overflowing of the Nile river each year. This is wrong! Egypt had water, that's true, but it was because of the pyramids that were constructed along the Nile river. Again, to blow the theory that pyramids were to entomb pharaohs is ludicrous, since there were many more pyramids built than pharaohs that had ever lived. Now, once again, this involves some physics to understand how the pyramid pumps water, but again, a 12th grade student will be able to comprehend how it works.

First of all, the reader must understand that in those days the earth's magnetic field was approximately 12 times the intensity that it is today. This can be verified from books in the library, and through modern research today, and is one of the most important aspects regarding how the pyramid pumps water.

Next, the earth's magnetic field is not a constant force, but one that has a pulsing action. This is very important, as you will see later.

Now that we have some understanding about the earth's magnetic field, as it was then, we will show how the pyramid utilizes this field to pump water year round, continually.

I feel the best approach to explain this would be to go through the construction of the "Great Pyramid."

First, the pyramid is constructed so that the base is aligned perfectly to the earth's north magnetic pole. Second, the construction of the north face of the pyramid is directly perpendicular to the earth's magnetic lines of force.

It is noted that Pyramids built in Africa to pump water have steeper slopes, since in this location, the earth's magnetic lines would be more parallel to the earth's surface. Pyramids in the United States (called mounds) have much less of a slope, since the earth's magnetic lines of force strike the earth in a more perpendicular direction.

What is the reason for this? The pyramids (or mounds) want to capture as much energy from the earth's magnetic lines of force as possible, similar to the way a sail on a sailboat will go faster (with more energy) when the sail is perpendicular to the wind. Wind is energy and the earth's magnetic field is energy. The larger the pyramid, the more energy it can capture. The pyramids have a very unique way of utilizing the earth's magnetic lines of force to pump water which will be shown later.

Next, the north surface of the pyramid must be constructed of a material that will be a conductor of the earth's magnetic lines of force. The earth's magnetic field is conducted by the use of "red granite". Iron particles within the granite make an excellent carrier for a magnetic field. Red granite was used in the construction of the "Great Pyramid," and it was brought from hundreds of miles away.

The next obvious question is where does this huge magnetic field lead that has been captured by the north side of the pyramid's surface. The north surface of the Great Pyramid is slightly concave from the outer edges to the center of the face. The earth's magnetic field follows the (path) generated by the iron particles located inside the red granite, bringing a very concentrated magnetic field into the so-called king's chamber.

Next, in the king's chamber, a large diameter, soft iron solid shaft, is placed. Around this iron cylinder, many coils of insulated copper wire are wound. The cylinder, with the insulated copper wire wound around it, is placed up against the exact location, where the red granite has brought all the earth's magnetic lines of force from the north pyramid's surface into the king's chamber.

What happens now is obvious to any student who has studied the operation of a simple transformer. To those who have not studied this principle, whenever a pulsating magnetic field passes through an iron core whose outer surface is wound with insulated copper wire, an electrical current will be generated within the insulated copper wire.

So, now the reader knows that, inside the king's chamber, is located a very large transformer that has a huge magnetic field pulsing through its iron core, which by electrical engineering laws, produces a very large, pulsating, current through its copper windings.

Before we get to what happens next, let us discuss where all pyramids are located. If you look at a map showing all the pyramids in the world, you will note that, at the time of their building, they were located near a river or built over a natural spring. Why? Because it is necessary to have a water source below the pyramid in order for the pyramid to pump water.

A good example is the Great Pyramid which I have used as a typical pyramid, it shows that there was an underground canal leading to the Nile river, and there have been reports, that in earlier years, people investigating the Great Pyramid, reported seeing water under it. Of course, the Nile river has changed its course over the thousands of years, and is not the same as when the Great Pyramid was built.

Now we have established two (2) things. 1. In the King's Chamber, a huge transformer uses the earth's magnetic lines of force to produce electrical power, and, 2. Water is below the pyramid.

Now, to the average 12th grade student, he has heard of, or seen in his chemistry lab, "ionized water." He knows then that ionized water is water that has ions (charged particles) inside the water which is charged by having one or more of the electrons either added or removed from the water molecules H_2O . In the case of the ionized water within the pyramid, it is always positively ionized, since it is always the electrons that are pulled away from the H_2O (water) molecules.

Now, to explain as simply as possible how the pyramid pumps water: The pyramid uses high charges of static electricity to pump huge quantities of water from the bottom central location below the pyramid up through a water tunnel to the outer surface. In the case of the Great Pyramid, it discharges water about 55 feet above the ground level. As an engineer, I can assure you, that a 55-foot head of water can force water a great distance on a flat area. This is like a huge water tank that is 55 feet above the ground, and it will have a large water pressure at ground level.

At this point, it would be easier to understand by referencing drawing Fig.1 to describe other parts of the pyramid, and what their function is, in order to understand how the pyramid pumps water. This drawings represents almost to scale all the items that exist inside the "Great Pyramid"

At the bottom of the pyramid, which we will call the "Lower Chamber," there is a water canal that brings water in from the Nile River. The water enters the ionization chamber, item 8, which contains two (2) electrodes. One is called the ionization grid electrode, Item 7, which is a wire grid (gold alloy), and the other is a Water Repulsion Electrode, with small orifices located on its top surface, Item 6.

Prior to the water entering the ionization chamber, the water flows through a trap, Item 5, that filters sediment from the river, and this is cleaned out periodically to prevent mud and sediment from clogging up the orifices, Item 27, of the Repulsion Electrode. The Ionization Chamber is not fixed, but may be adjusted so that the water level of the Nile river will always be at the top of the Ionization Chamber. Therefore, the ionization chamber will always be filled with water, unless the river makes a radical change in direction.

To explain how this electrical process works as simply as possible: Inside the so-called Queen's Chamber is a very large capacitor, Item 21, which stores huge amounts of positive electrical charges from the electricity generated by the transformer located in the so-called King's Chamber. Electrical charges flow from the Capacitor, then to the Electrical Distributor, Item 22, whose contacts rotates at a rapid rate, and sends high voltage positive charges of electricity to the Ionization Chamber, Item 8. There are two (2) insulated wire conductors that are connected to the Ionization Chamber. The first wire conductor, Item 28, is connected to Ionization Grid Electrode, Item 7, and the second wire conductor, Item 24, is connected to the Water Repulsion Electrode, Item 6.

Inside the Ionization Chamber, the Ionization Grid Electrode is made up of many high voltage ionization wires, which pulls off huge amounts of electrons from the H_2O molecules when energized by the highly positively charged wires, Item 28.

Immediately below the Ionization Grid Electrode is the Water Repulsion electrode, Item 6, which is connected to the positively charged wire conductor, Item 24. The top surface contains many small orifices, Item 27, which is in the form of a box with the bottom side of the box open to the Nile river water.

Due to the configuration of this electrode, as shown on figure 2, when this electrode is positively charged, all the positive charges are forced to the outer surface of this electrode. This is a proven phenomenon, and has been demonstrated in our Physics and is known as "Hook's Law."

As the water passed up through the Water Repulsion Electrode, Item 6, and through the small orifices, Item 27, the positively charged Ionization Grid Electrode, Item 7, ionizes the water to highly positively charged ionized water, which is immediately above Item 6.

The Ionization Grid Electrode is de-energized, and before the positively charged ionized water has any time, whatsoever to re-establish itself into homogenously electrically neutral water, the Water Repulsion Electrode Item 6 is instantly energized with a huge positive charge by the electrical conductor, Item 24, and repels the positively charged H_2O molecules through the uncharged, Ionization Grid Electrode Item 7, and then up through the approximately 4 feet x 4 feet water discharge shaft, Item 9, of the pyramid. It spills over the side of the pyramid at approximately 55 feet above ground level. It spills over to a water trough, Item 13, that leads to the causeway in front of the pyramid.

This process repeats itself at a very rapid rate, due to the very fast alternating pulsing action generated by the rotating Electrical Distributor, Item 22. It noted that an electric field can be established in 1×10^{-17} second.

As the high voltage electrons Item 25 are removed from the water during the process of ionization, they are emitted from the pyramid by the Electron Emission Rod, Item 17, at the apex of the pyramid. At the apex where the electrons are emitted, there will be a reddish orange glow at night, surrounding this Electron Emission Rod, as the high voltage negatively charged electricity ionizes the air around it. This is known as "electrical corona."

The Ionization grid is about 1/2 inch above the Water Repulsion Electrode where the water is ionized, and again this goes back to Physics, where we can see how tremendous repulsive forces can be generated on the ionized water when the "same charges" are brought in close proximity of each other. It is the following application of Coulomb's Law that allows the pyramid to generate the necessary forces to pump the water out of the pyramid, which is similar to the propulsion of the so-called "Flying Saucers" See patent in back of book for similarity:

Force = $(9 \times 10^9) Q_1 \times Q_2 / R^2$ = newton

Force = force acting on the ionized water molecules to repel up the shaft

(9×10^9) = constant which is "speed of light squared"

Q_1 = charge on the propulsion electrode (coulomb)

Q_2 = charge on the ionized water (coulomb)

R^2 = Distance (meters) between the charges (squared) distance between positively charged electrode, Item 6, and the positively charged ionized water immediately above it.

It is noted that inside the Ionization Chamber, above the Water Repulsion Electrode, Item 6, the chamber is water tight, except for the opening leading to the water shaft. Each time the Water Repulsion Electrode is energized, the force of the repulsion action, within the chamber, would be synonymous to the manner in which the force of a piston would act inside the cylinder of a water pump. Since the Ionization Chamber is water tight, the only direction that the pressurized water can take is upward through the pyramid's water shaft, Item 9.

As the water is pumped upward through the water shaft of the pyramid, it is possible to store huge quantities of water to the "Great Gallery", Item 11. The so-called "Great Gallery" is such a water storage tank for water to be use later. There is located a 2-position water valve (stone blocks) Item 10, that can be moved to shut off water flow to the portion of the water shaft that leads to the outer surface of the pyramid, and allows the water to flow upward to fill the "Great Gallery" with stored water.

When immediate water is required by the population, the 2-way stone valve is repositioned, which allows the water to flow from the "Great Gallery" into the water shaft, then to the outside surface of the pyramid into the water trough, Item 13,

Incidentally, the stone 2-way valves are currently there, located inside the" Great Pyramid," as well as mud sediment from the Nile River, which is still on the walls of the

"Great Gallery" and inside the walls of the water shaft, that have remained there over the thousands of years.

From Figure 1, it shows the two (2) shafts, Item 15, (as shown on TV many times), which are located in the so-called "King's Chamber."

The reason for the shafts being there is: As the water rises inside the "Great Gallery" to store water, if air vents were not present, the internal air pressure within the pyramid would build up as water entered the "Great Gallery" and it would be impossible to fill the gallery with water.

Also, when the water is released from the gallery, if vents were not present, a partial vacuum that would be created above the water level would prevent water from leaving the gallery. Therefore, air vents to the outside atmosphere must be above the water level to allow both entering and leaving of water from the so-called "Great Gallery". There are two shafts located there, in case one of the vent shaft became "clogged" for any reason, and the water supply could still be delivered to the population.

In the days when the Great Pyramid was in operation, great amounts of water would flow from the Great Pyramid into the area around the "Sphinx." There still exists a brick wall around the "Sphinx" which, at that time, was filled with water, thus forming a man-made lake, or moat, around the "Sphinx".

The "Sphinx" was a water distribution point, with several water canals, allowing water to flow to various areas of the countryside.

 It is noted that in 1998, archeologists have discovered that tunnels indeed do lead out form the "Sphinx" to the countryside. It is my prediction that they will find openings in the top wall of these tributaries, so that the people could draw water from this underground water supply for domestic purposes.

It is also noted that, as water flowed from the pyramid down the causeway, spilling over the wall into the "Sphinx" area, that water eroded the wall next to the causeway.

For many years archeologist thought it was "wind" erosion, but now they have concluded it was definitely <u>WATER EROSION</u> that caused it. This alone proves that the water came from the pyramid, down the causeway and spilled over into the "Sphinx" enclosure.

Archeologist are very confused now, since they know from weather data that there was not enough rain water in that day to have caused water to erode the wall.

Also noted in the past, copper wire has been found inside the pyramid which would be necessary for the transmission of electrical power required for the ionization, and propulsion of the water. Of course, no one knew what the copper wire was used for inside the pyramid, but now, its use is described in the book.

It is noted, that the larger that pyramid is built, the greater its capacity to pump water. This would be obvious by its ability to capture more of the earth's magnetic lines of force, thus generating greater electrical power. The more electrical power generated, then the greater the electrical power to ionize the water and for repulsion of the water.

The Great Pyramid has the capacity to pump water indefinitely, provided there was water in the vicinity, and the energy of the earth's magnetic field stayed at the strength as it was then. The pyramid contains no moving parts to wear out, except for an electrical distributor. For the people of Egypt, this was a wonderful structure in the desert, perpetually pumping liquid gold (water) for domestic purposes as well as irrigation of the crops.

CHAPTER 2 - EMBOIDMENTS OF THE DISCOVERY

I will now identify the items shown on figures1 and 2, which will aid the reader in understanding how all the parts fit together, in order for the pyramid to pump water. Most of these items currently exist on the Great Pyramid of Giza.

Item 1 - This is the Nile River which supplies the water to the pyramid.

Item 2 - This is the entrance to the water tunnel that accesses the pyramid.

Item 3 - This is the water tunnel that allows water to flow under the pyramid (existing today).

Item 4 - This is a clean-out in the tunnel to remove any mud or debris before water enters the Ionization Chamber.

Item 5 - This is a water trap that allows any mud or sediment to collect before entering the Ionization Chamber and clogging up the small orifices within the Water Repulsion Electrode, Item 6.

Item 6 - This is the Water Repulsion Electrode. The top surface of this "box like" electrode is perforated on the top, and when energized with positive electricity, it repels the water upward through the water shaft, Item 9.

Item 7 - This is the Ionization Grid Electrode, which when energized with positive charged electricity, it removes huge amounts of electrons from the water, which ionizes the water to a high state of positive ionization.

Item 8 - This is the Ionization Chamber which is constructed of a high strength dielectric material, and having a high dielectric constant, which houses the Water Repulsion Electrode, Item 6, and the Ionization Grid Electrode, Item 7.

Item 9 - This is the Water Shaft where the water travels through to get either to the "Great Gallery" (water storage) or to the outside trough of the pyramid.

Item 10- These are the Rock Valves (still existing) that are movable to allow water to either flow to the "Great Gallery" or to flow to the outside edge of the pyramid.

Item 11 - This is the so called "Great Gallery" which is nothing more than a huge water storage tank which is elevated about 60 feet above ground level.

Item 12 - This is the water discharging over the side of the pyramid into the trough which will give the water a head pressure of about 55 feet above the ground elevation at the bottom of the Great Pyramid. This head pressure can carry the water a great distance from the pyramid.

Item 13- This is the water trough that the water spills into on the side of the Great Pyramid, and discharges into a water aqueduct (or causeway) that discharges the water over the wall, and into the enclosure surrounding the "Sphinx".

Item 14- This is the Red Granite face of the Pyramid that channels the flow of the earth's magnetic lines of force into the so-called "King's Chamber."

Item 15- These are the air vents that allow air to escape from the so-called "Great Gallery" as water is filling the gallery, and also allow air to enter the gallery as water is being released from the gallery. Water could not leave the gallery if air could not come in through these vents. If these vents were not there, the "Great Gallery" (storage tank) would become pressurized and water could not flow into it.

Item 16- These are the earth's magnetic lines of force as they strike the face of the pyramid at a perpendicular angle, which is 90 degrees to the face of the pyramid.

Item 17- This is the Electron Emission Rod. It is a gold alloy, sharp pointed negatively charged electrode, that is attached to the top of the pyramid. When energized by the high voltage generated by transformer copper windings around the soft ion core, draw the electrons from the water, and under high voltage, emits huge amounts of electrons into the atmosphere. This would have a "reddish-orange" glow at night denoting a highly negatively charged electrode .

Item 18- This is the so-called "King's Chamber" where the soft iron core cylinder wrapped with copper windings (electrical transformer) are located to produce electricity from the earth's "pulsating" magnetic field travelling through the ferrous granite of the north face of the pyramid, and into the soft iron core.

Item 19- This is the soft iron core located in the so-called "King's Chamber" where all the pyramid's magnetic flux line from the northern face of the pyramid are concentrated and flows through in a pulsating manner. Pyramids of different sizes would have different sizes of iron cores.

Item 20- These are the copper (insulated windings) that surround the soft iron core. It is noted in our laws of electrical physics, that the more winding that placed around the iron core, the greater the voltage produced to the Ionization Chamber, Item 8.

Item 21- This is the System Capacitor. This a very large capacitor which acts similar to an electrical storage tank, and stores huge amounts of electrical charges furnished by the electrical transformer, Item 25, necessary to furnish high voltage positively charged electricity to the Ionization Chamber, Item 8.

Item 22- This is the Electrical Distributor. It is composed of two (2) rotating disks that alternately distributes high voltage electricity to both the Ionization Grid Electrode, Item 7, and the Propulsion Electrode, Item 6.

Item 23- These are electrons that are being removed from the water and are traveling from the Ionization electrode, Item 7, to the capacitor, Item 21, and to the Emission Rod, Item 17, located at the apex of the pyramid.

Item 24- This is the positively charged electrical conductor that connects the electrical distributor to the Water Repulsion Electrode.

Item 25- These are billions upon billions of electrons that are emitted to the atmosphere by the negatively charged high voltage electrode, Item 17. A reddish, orange "corona" will be seen from this electrode when the pyramid is operating at night.

Item 26- These are electrical diodes that allow positive and/or negative electricity to flow in one direction only. It is these electrical devices that allow only positive electricity to flow into the Ionization Chamber.

Item 27- These are the small orifices that are located on the top surface of the Water Repulsion Electrode to allow water to pass through to the top side. There the water is repelled away from this electrode, and up the water discharge tunnel.

Item 28- This is the positive charged electrical conductor that connects the Electrical Distributor to the Ionization Grid Electrode.

Item 29- This is the conductor that carries electrical charges to the electron emission rod, Item 17

I reiterate the statements that I have made earlier. There is a great debate going on today that exists between the historians and the geologists. The argument (discussion) can be seen quite often on the Learning Channel of your TV. As mentioned earlier, geologists have found that it was not wind that eroded the side walls of the "SPHINX."

adjacent to the wall of the "Causeway" leading to the Great Pyramid, but WATER. As discussed, it was water coming from the pyramid, and flowing down the causeway, spilling over the wall and eroding it over a long period of time.

Our historians think the Pharaohs built the pyramids about 5000 -6000 BC. This is not correct, since for the pyramids to pump water, the magnetic field of the earth would have to be approximately 12 times its strength as it exists today, which would put the pyramid's construction at about 10,000 to 12,000 years ago.

In other words the pyramids existed several thousand years before the Pharaohs came on the scene. The magnetic field of the earth, at the time of the pharaohs, would not have been strong enough to produce enough electricity to pump the water up the shaft of the Great Pyramid.

Also, our historians have very accurately determined that there was much rainfall in the Gaza area during earlier times in our past history. The only problem is, and the hole in their hypothesis is the fact that there was approximately 10,000 years between the rainfall, when they think the "GREAT PYRAMID" and the "SPHINX" were built. This conflicts with the fact that there is actual physical evidence that water caused the erosion around the wall of the "Sphinx" leading to the pyramid, that exists today.

To reiterate: The water traveled from the Great Pyramid, down the existing "Water Causeway," and spilled over the "water containing wall" surrounding the "SPHINX" and eroded the inside of the wall, as the water formed a **LAKE**, or moat, surrounding the "SPHINX". Also, should anyone investigate the stone floor surrounding the "SPHINX", they will find that it is virtually water tight in order to keep the water from leaking out through the sandy soil below.

As mentioned, from the "SPHINX" location, archeologist have found various underground tunnels. These underground tunnels were used to distribute water to the populated areas for the purpose of domestic drinking water, washing clothes, and the irrigation of the crops.

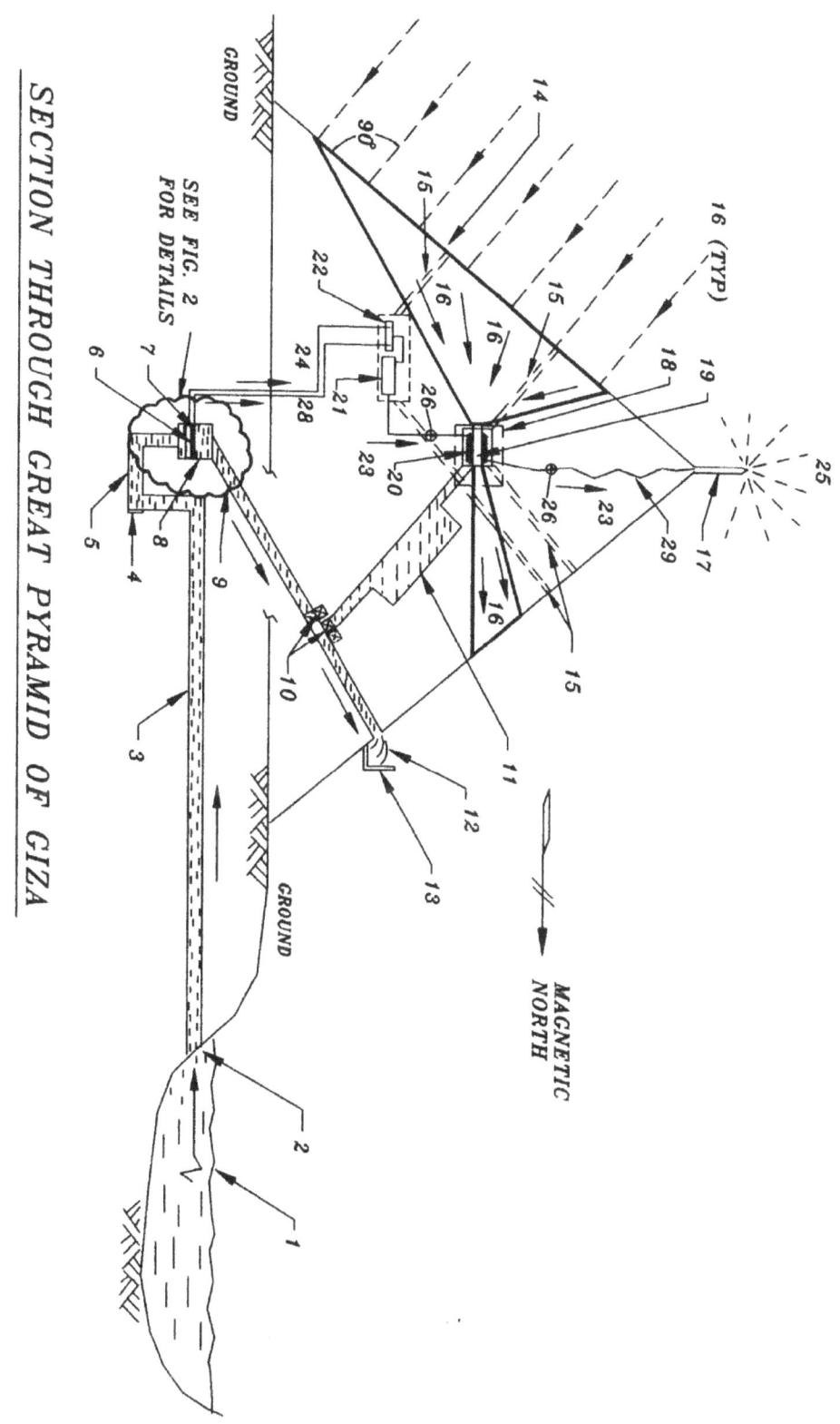

SECTION THROUGH GREAT PYRAMID OF GIZA

FIG.1

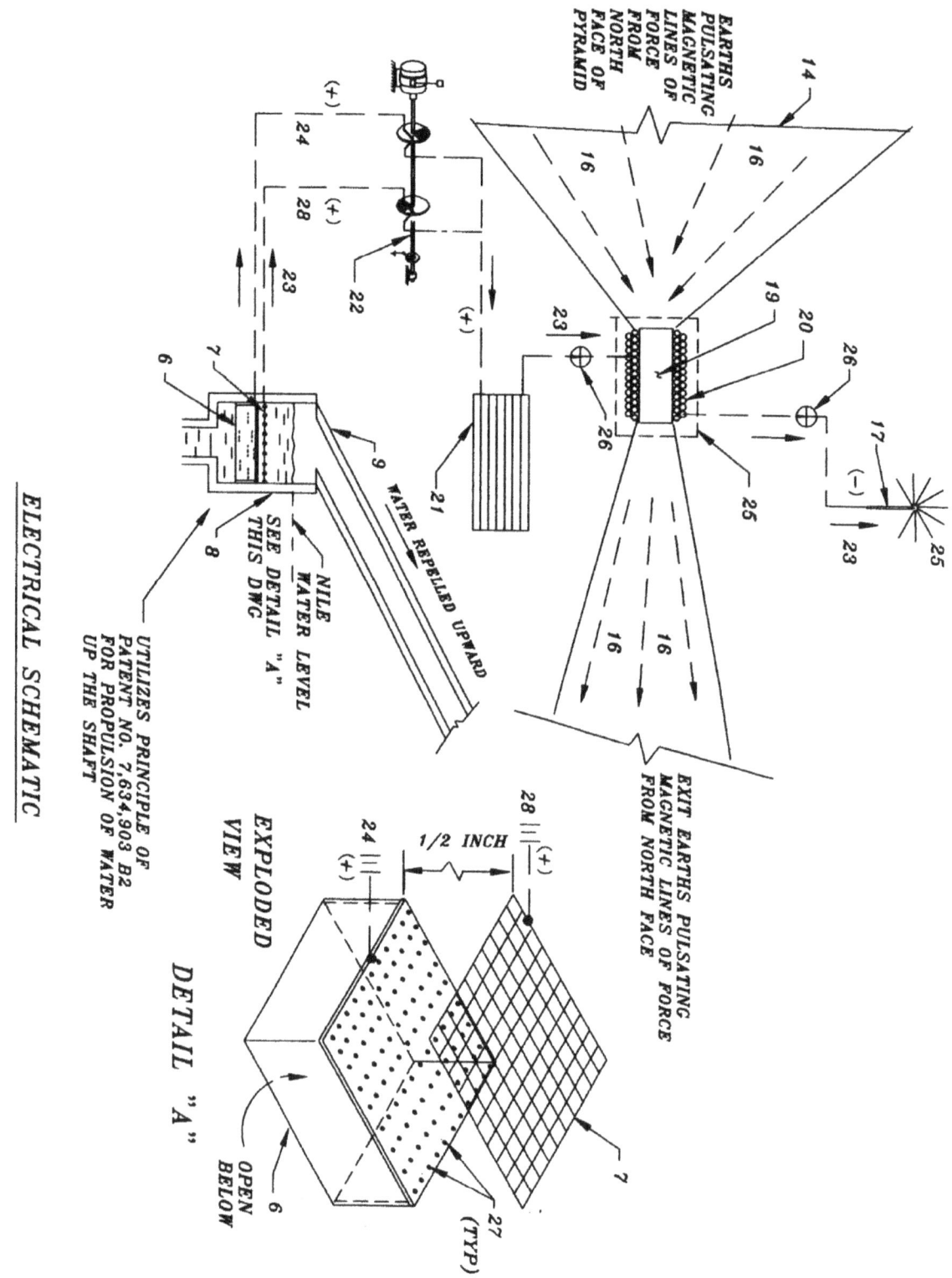

ELECTRICAL SCHEMATIC

EARTHS
PULSATING
MAGNETIC
LINES OF
FORCE
FROM
NORTH
FACE OF
PYRAMID

EXIT EARTHS PULSATING
MAGNETIC LINES OF FORCE
FROM NORTH FACE

WATER REPELLED UPWARD

NILE
WATER LEVEL

SEE DETAIL "A"
THIS DWG

UTILIZES PRINCIPLE OF
PATENT NO. 7,634,903 B2
FOR PROPULSION OF WATER
UP THE SHAFT

EXPLODED
VIEW

DETAIL "A"

1/2 INCH

OPEN
BELOW

(TYP)

FIG.2

The United States of America

The Commissioner of Patents and Trademarks

Has received an application for a patent for a new and useful invention. The title and description of the invention are enclosed. The requirements of law have been complied with, and it has been determined that a patent on the invention shall be granted under the law.

Therefore, this

United States Patent

Grants to the person or persons having title to this patent the right to exclude others from making, using or selling the invention throughout the United States of America for the term of seventeen years from the date of this patent, subject to the payment of maintenance fees as provided by law.

Harry F. Manbeck, Jr.

Commissioner of Patents and Trademarks

Attest

(12) **United States Patent**

Phillips

(10) **Patent No.:** **US 7,634,903 B2**

(45) **Date of Patent:** **Dec. 22, 2009**

(54) **ION IMPULSE ENGINE**

(76) Inventor: **Richard Cary Phillips**, 5458 Somersby Pkwy., Pinson, AL (US) 35126

(*) Notice: Subject to any disclaimer, the term of this patent is extended or adjusted under 35 U.S.C. 154(b) by 624 days.

(21) Appl. No.: **11/247,087**

(22) Filed: **Oct. 12, 2005**

(65) **Prior Publication Data**

US 2007/0079595 A1   Apr. 12, 2007

(51) **Int. Cl.**
F03H 1/00   (2006.01)
B63H 11/00   (2006.01)

(52) **U.S. Cl.** **60/202**; 60/204

(58) **Field of Classification Search** 60/202, 60/204
See application file for complete search history.

(56) **References Cited**

U.S. PATENT DOCUMENTS

2,952,970 A * 9/1960 Blackman 60/202

3,050,652	A	*	8/1962	Baldwin	313/359.1
3,052,088	A	*	9/1962	Davis et al.	60/202
3,156,090	A	*	11/1964	Kaufman	60/202
3,304,718	A	*	2/1967	Webb	60/202
3,501,376	A	*	3/1970	Dow et al.	376/144
3,535,586	A	*	10/1970	Sabol	315/111.61
5,005,361	A	*	4/1991	Phillips	60/671
7,096,660	B2	*	8/2006	Keady	60/203.1
2005/0257515	A1	*	11/2005	Song	60/202

* cited by examiner

Primary Examiner—Michael Cuff
Assistant Examiner—Gerald L Sung

(57)   **ABSTRACT**

An ion engine which produces thrust by charging a propellant vapor in an electric field and then reversing the polarity of the electric field. The electric field forces the charged nuclei close to the unenergized electrode. When the electrode is energized as a positive electrical charge, the nuclei repel and create thrust in the opposite direction.

4 Claims, 4 Drawing Sheets

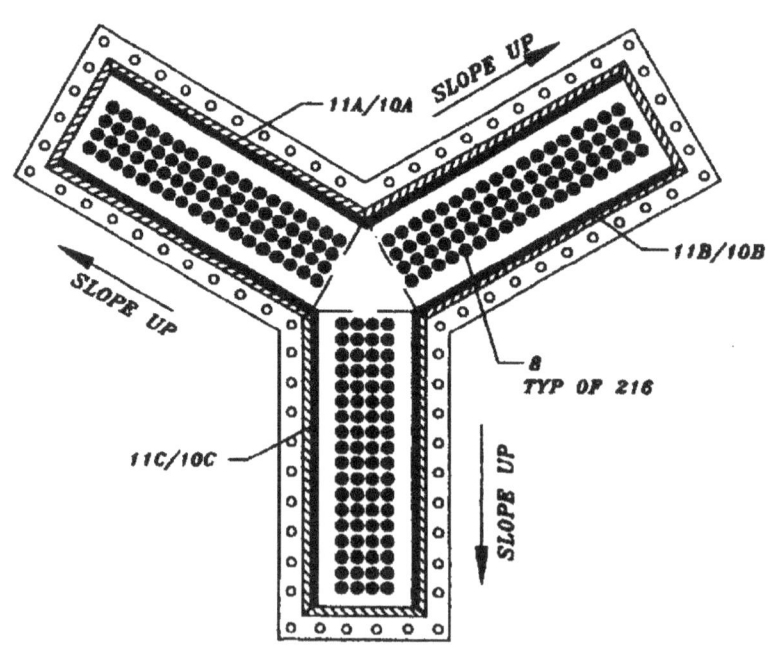

ENLARGED VIEW OF BOTTOM OF ENGINE

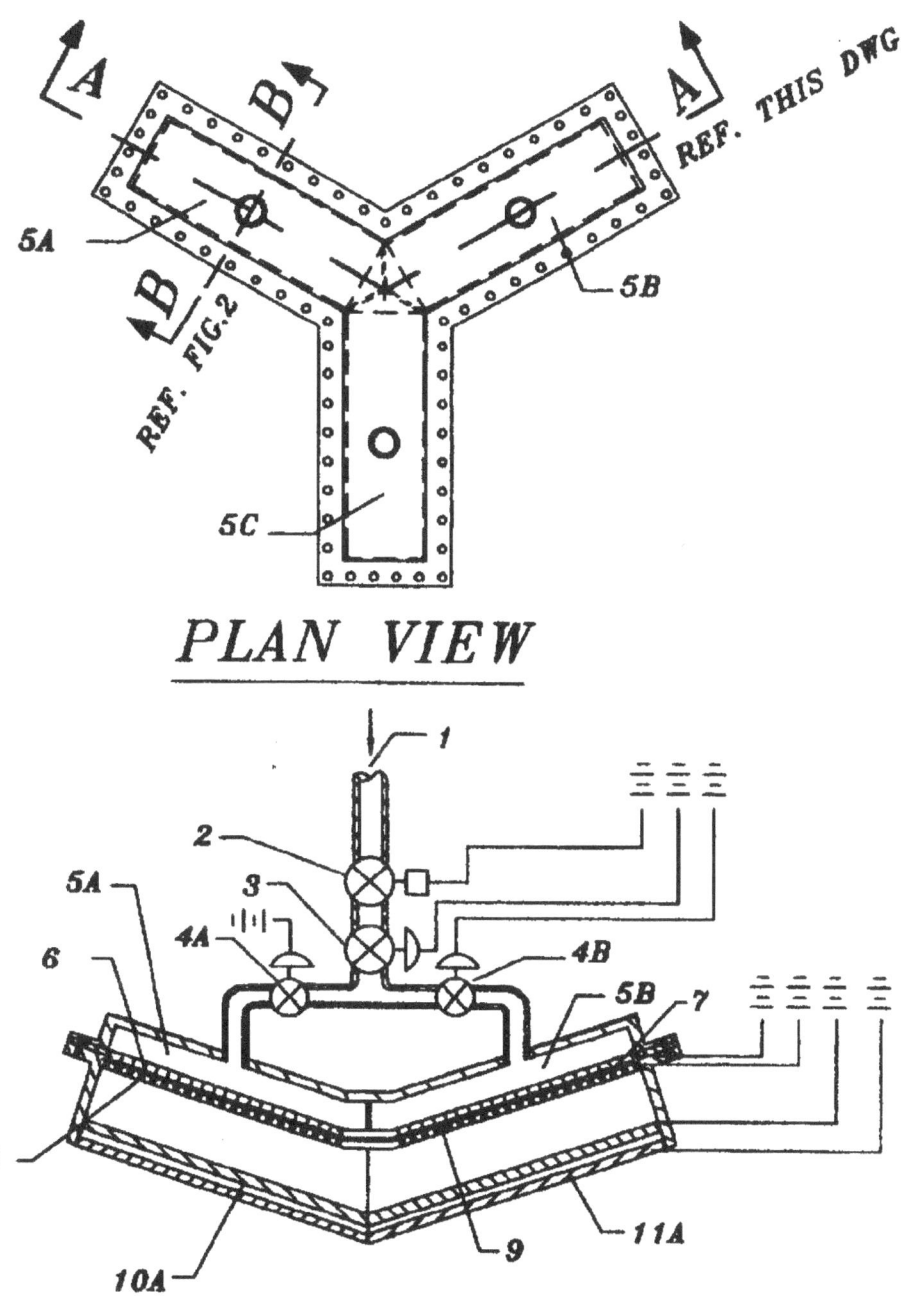

PLAN VIEW

SEC. A—A (REF. THIS DWG)

FIG.1

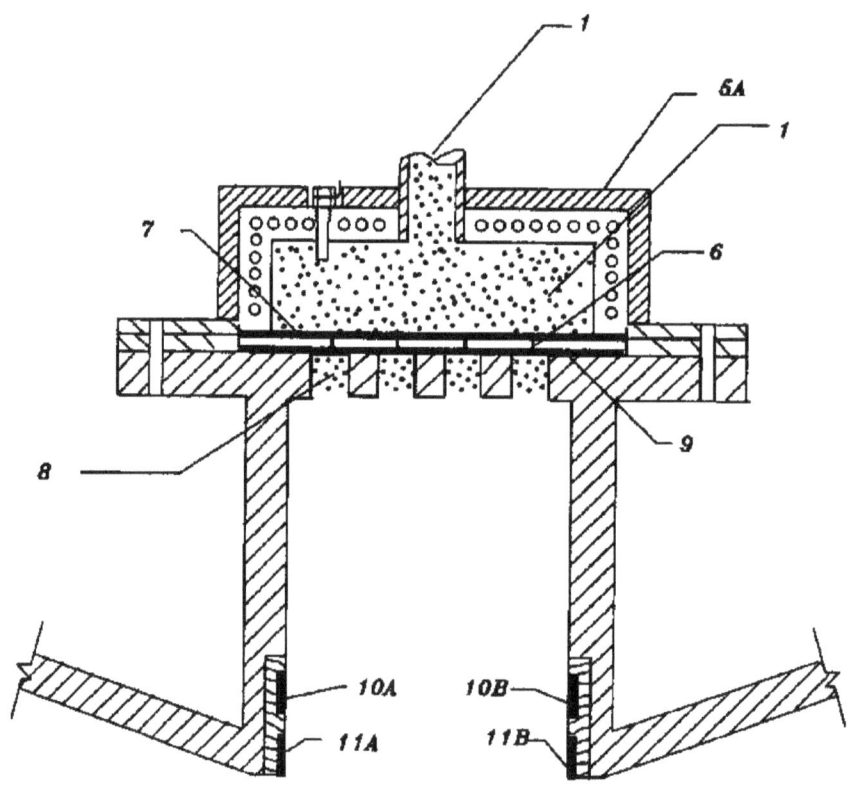

$$SEC. \ \ B-B \ \ \textit{(REF. FIG.1)}$$

FIG.2

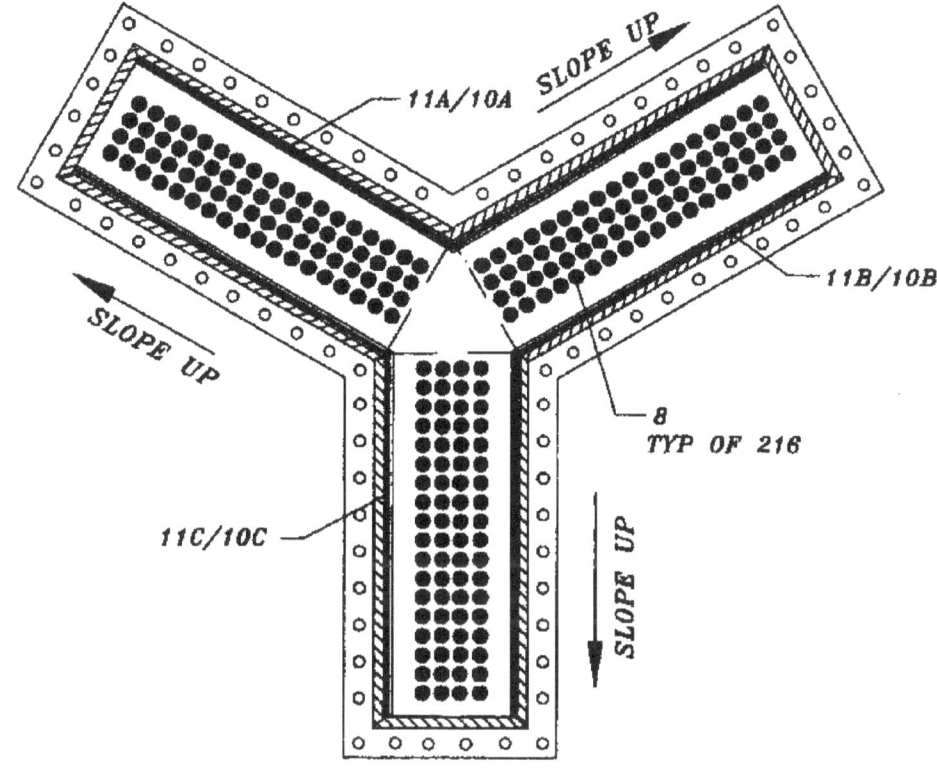

11A/10A

SLOPE UP

11B/10B

SLOPE UP

8
TYP OF 216

11C/10C

SLOPE UP

ENLARGED VIEW OF
BOTTOM OF ENGINE

FIG.3

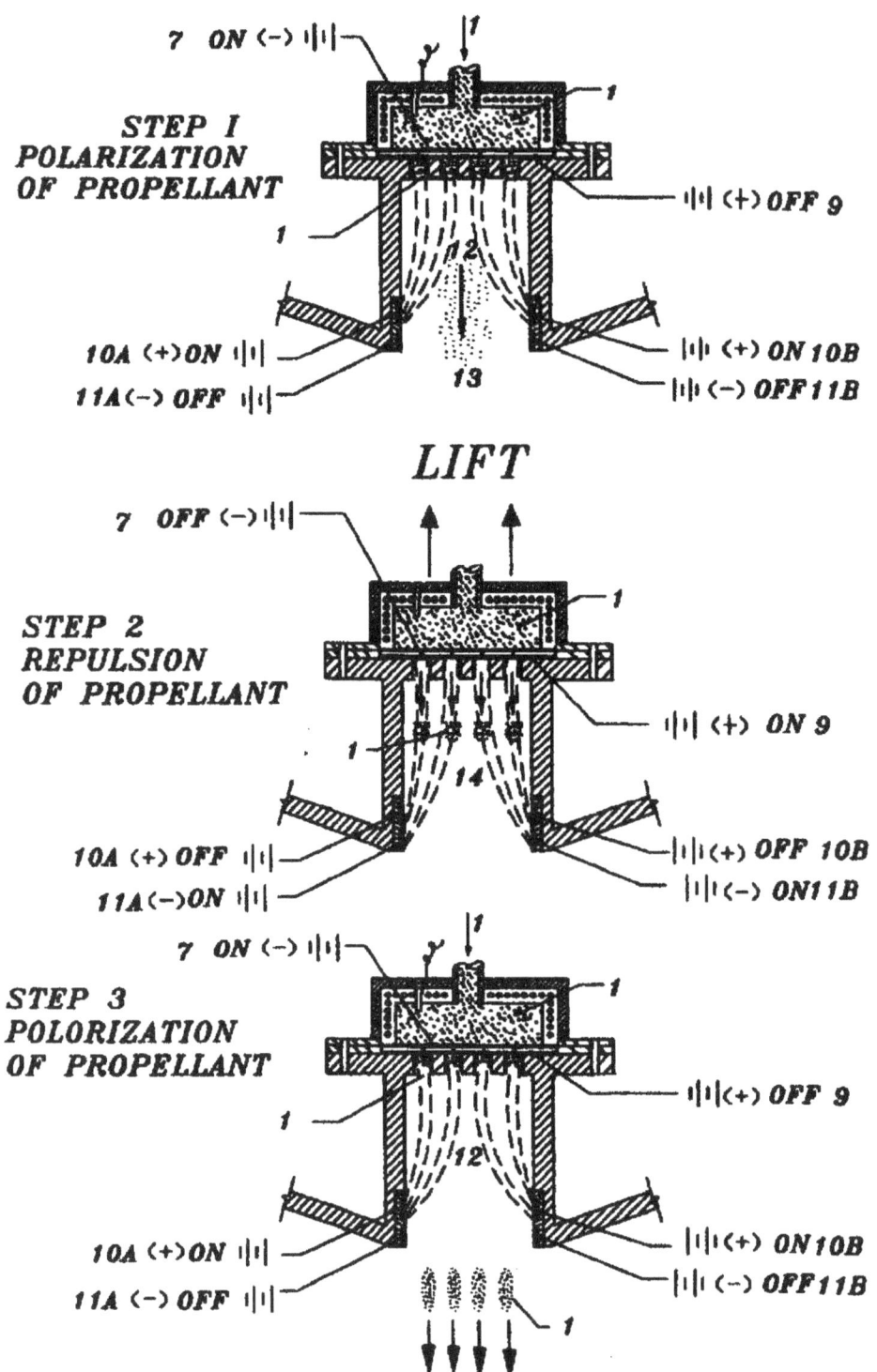

FIG.4   DIAGRAMMATIC PROCESS OF
LIFT GENERATED ON CRAFT

1

ION IMPULSE ENGINE

This application uses concepts from U.S. Pat. No. 5,005,361 issued on Apr. 9, 1991, to Richard C. Phillips to a new, useful, and different purpose.

The government has not sponsored any research or development monies of other aids in regard to this application.

BACKGROUND OF THE INVENTION

a. Technical Field

Drawn to a liquid/vapor ion power plant.

b. Description of The Prior Art

Currently there are very few useful applications of ions to generate thrust. Current methods produce very low thrust. The Phillips '361 patent uses ions to generate a rotating shaft with a turbine. The present invention eliminates the turbine and associated structure and uses the pure thrust generated by the ions to propel the engine. This reduces the mechanical loss inherent in transferring the thrust to the turbine and provides a superior method of propulsion for airborne, waterborne, or space traveling craft. The improved product significantly improves upon the existing state of the art and advances the use of ions to produce an engine to a practical, economical reality.

In addition to improving the ion to a propulsion engine, the present invention utilizes a very highly reactive vapor such as P_4O_6 which is much easier to polarize than mercury vapor when in the influence of an electric field, cost less, and has fewer environmental hazards. Sodium, zinc, and calcium based compounds are also effective. With a different construction for the polarization electrode, a different construction for the power electrode, a different ion discharge chamber configuration, and the addition of a negative electrode to establish the electric field that separates the ions, the present invention is superior to the Phillips '361 patent in virtually every way.

The engine can generate enough power to lift an airplane or spacecraft off the earth's surface and propel it in any direction. As no external oxygen is necessary for the operation of this engine, it can operate outside the atmosphere for spacecraft in space.

SUMMARY

This is an ion engine which utilizes a form of ion propulsion to generate power from a material propellant. The propellant can be any compound or element that can be vaporized and have its nuclei charged in an electric field. Some compounds and elements are better than others at holding a positive charge on the nucleus of the component atoms. The preferred propellant is P_4O_6, a man made compound that becomes a highly reactive vapor at a reasonably low temperature.

Thrust is derived by the construction of an ion repulsion discharge chamber, and the placement of positive and negative conducting electrodes, which act upon a net positive charge of the nucleus or nuclei, within the propellant vapors. Propellant vapor is delivered to a propellant entry chamber.

The basic elements of the engine include a housing; an ion discharge chamber divided into a plurality of sections, each section disposed at an angle from the other sections and each section open at a bottom end, sufficient to allow the escape of a vaporized material and adapted to connect to a plurality of propellant entry chambers, three sections angled on a slope in three dimensional space are optimum; a plurality of propellant entry chambers in the housing each open at the end

2

connected to the ion discharge chamber and adapted to receive entry of a vaporized material; a first positively charged electrode attached to the housing near the bottom end of the ion discharge chamber; a second positively charged electrode attached to the housing near the bottom end of the ion discharge chamber; a first negatively charged electrode attached to the housing between the first positively charged electrode and the bottom end of the ion discharge chamber; a second negatively charged electrode attached to the housing disposed relative to the second positively charged electrode such that the second negatively charged electrode is between the second negatively charged electrode and the bottom end of the ion discharge chamber; a third positively charged electrode attached to the housing near the area where the propellant entry chambers are adapted to receive the vaporized material; a third negatively charged electrode attached to the housing near the area where the propellant entry chambers are adapted to receive the vaporized material disposed close to the third positively charged electrode and with the third positively charged electrode between the third negatively charged electrode and each propellant entry chamber; a means to control the time at which the electrodes are energized and de-energized.

Briefly stated, a vaporized propellant material is metered and injected into a propellant entry chamber. The propellant can be stored in tanks in liquid or solid form and subsequently heated to vapor form or it can be stored in vapor form if desired. Pipes or tubes carry the pressurized propellant from the storage container to the propellant entry chamber. One or more valves may be used to regulate the flow of propellant. The propellant material flows from the propellant entry chamber to the ion discharge chamber. The propellant vapor is acted upon by a charged electric field that is termed a polarization field. Positive electrodes at the open end of the discharge chamber and a negative electrode behind the ion entry chamber create this polarization field. The polarization field causes the electrons of the atoms to separate from the nuclei creating positive charged ions and free negatively charged electrons. This polarization field also creates a separation between the net charge of the positive ions and the net charge of the negative electrons within the propellant vapor. The separation of the net charges is accomplished at a very fast rate of speed in a strong electric field. The separation physically moves the positive nuclei close to the negatively charged electrode at the propellant entry chamber.

The polarization field is then turned off by de-energizing the electrodes used to create it, and simultaneously another high voltage electric field is established by energizing the positively charged electrode on the end of the propellant entry chamber and the negative electrodes at the end of the ion discharge chamber which acts upon the net positive charge within the vapor. This field is termed the power field because it generates the thrust. The positively charged electrode is very close to the net positive charge of the nuclei of the propellant vapors, and repels the nuclei of the propellant vapors (positive ions) out of the open end of the ion discharge chamber at an extremely high velocity. The engine and anything attached to it is thrust in the opposite direction due to the equal and opposite reaction force generated, as the net positive charge of the ions, within the of the propellant vapor, are repelled. The power field is then turned off, more propellant is pumped into the ion entry chamber and the process is repeated. This can be done a number of times per second limited only by the amount of propellant available. It can be done as fast as the vapor valves can operate with pressurized vapor, as an electric field can be generated in a time span of

1×10^{-17} seconds. Each time the polarization field is interrupted by turning it off and the power field established, a pulse of thrust is generated.

Additional elements include means to supply the vaporized material, means to power the electrodes, means to control the flow of the vaporized material, and a means to time the energerization of the electrodes. Pipes, valves, and manifolds may be used to supply the vaporized material to the propellant entry chambers; electric switches time activated provide a means to time the energization and de-energization of the electrodes such that the first and second positive electrodes and the third negative electrode are energized while the first and second negative electrodes and the third positive electrode are not energized. At specified times these electrodes are de-energized and the first and second negative electrodes and the third positive electrode are energized. Electrically powered valves may be used to regulate the flow of propellant into the propellant entry chambers. On off switches are used as a means to energize and de-energize the electrodes. Electric power may be provided by batteries.

This invention utilizes "Charles Coulombs Law" generating huge electro-static forces when electrical charges of the same "electrical sign" are in close proximity of each other. By electrically forcing the positive ions to a position very close to the unenergized positive electrode plate, the force generated when the positive electrode plate is energized, creating the positive electrical charge, is increased over the force that would be generated by not electrically forcing the positive ions into close proximity to the positive electrode plate.

The following formula by Charles Coulomb is utilized by my invention, and is very important to note the following in this formula: "R" represents the distance between the net positive charge of the vapors and the positive charge on the power electrode (both of the same electrical sign) and is located in the denominator of this formula. In my invention, by the use of a special polarization field inside the ion discharge chamber, it acts on the propellant vapors and locates the net positive charge of propellant vapors extremely close to the positively charged electrode. Therefore, in my invention, the "R" (distance) in Charles Coulombs force formula represents a very short distance when the vapors are repelled out of the chamber.

Also, noted by the use of mathematics in analyzing Charles Coulombs force formula, that as the "R" (distance) between these charges approach "Zero" (0), which is in the denominator of this formula, then the force generated by my engine approaches infinity. Charles Coulombs formula is as follows:

$$\text{Reaction Force} = \frac{[9\times10^9]\,\|Q_1\|\,\|Q_2\|}{R^2}\text{Newtons}$$

where:

Q_1=Net positive charge of the ions (coulombs)
Q_2=Positive charge on the (power) electrode (coulombs)
R^2=Distance (meter) between the net positive charge of vapors and the positive charge on the power electrode
$[9\times10^9]$=Constant involving the speed of light (squared)
Force=Newtons

The practical engine pulse rate is designed for the electrical fields to pulse at the rate of 1.5 times per second; however, fields can be pulsed at much faster rates or at slower rates if desired merely by regulating the time that the polarization field is turned off and on. Obviously, faster pulse rates of the vapors will generate greater thrust from the engine, since the time span between each pulse is a function of power.

One very important variant feature of this engine, which is different from the prior art, is the arrangement of the ion discharge-chambers into 3 groups. Where two or more, preferably three (3) sections are used, this allows it to produce a thrust force on the engine in any direction. See FIGS. 1, 2, and 3 for details of this arrangement. This allows the movement vector to be adjusted in any direction in three dimensional space.

BRIEF DESCRIPTION OF DRAWINGS:

FIG. 1 is showing plan view of engine and Sec. A-A is showing the cross section of the electrodes within the ion repulsion discharge chamber.

FIG. 2 is Section B-B showing the cross section of the ion repulsion discharge chamber.

FIG. 3 is an enlarged view of the bottom of the engine.

FIG. 4 is a diagrammatic process describing how lift is generated on the engine.

DETAILED DESCRIPTION OF THE PREFERRED EMBODIMENTS

The purpose of this engine is to produce power from a high temperature vapor of P_4O_6, phosphorus trioxide, and high voltage electricity. P_4O_6 is made from fertilizer and falls harmlessly to the earth, unlike combustion products, of rockets. P_4O_6 decomposes into other environmentally less hazardous products than chemical fuels, solid fuels, mercury or other propellants, consequently, use of P_4O_6 is both the most effective propellant and the least harmful.

Under normal conditions, P_4O_6 melts at 72.5 degree F. and flows from the propellant fuel tanks through solenoid (on-off) valves and then into the propellant vaporization chamber where it boils at 343 degrees F. and is transformed into a vapor. The vaporized P_4O_6 (1) then flows through the propellant system shut off valve (2), then flows through the propellant throttle flow control valve (3), then flows through any combination of the propellant directional valves (4a,4b,4c) which then flows into three propellant distribution-manifolds (5a,5b,5c). From the three propellant distribution manifold the vaporized P_4O_6 flows through small orifices (6) into 216 cylinders referred to as propellant entry chambers (8), each approximately 1 inch diameterx^1 ½ inch deep, all located on the upper end of the ion discharge chamber.

The polarization field will be created by using three positively charged electrodes and three negatively charged electrodes. There will be a first positively charged electrode attached to the housing near the bottom end of the ion discharge chamber, a second positively charged electrode attached to the housing near the bottom end of the ion discharge chamber and directly opposite the first positively charged electrode, and a third positively charged electrode attached to the housing near the area where the propellant entry chambers are adapted to receive vaporized propellant. There will be a first negatively charged electrode attached to the housing between the first positively charged electrode and the bottom end of the ion discharge chamber, a second negatively charged electrode attached to the housing opposite the first negatively charged electrode near the bottom end of the ion discharge chamber and disposed relative to the second positively charged electrode such that the second negatively charged electrode is between the second positively charged electrode and the bottom end of the ion discharge chamber, and a third negatively charged electrode attached to the housing near the area where the propellant entry chambers are adapted to receive vaporized propellant disposed close to the third positively charged electrode and with the third positively charged electrode between the third negatively charged and each propellant entry chamber. The ion discharge chamber is adapted at the end near where the third positively charged electrode is attached to the housing connect to a plurality of propellant entry chambers. This will generate the polarization field necessary for the engine to function.

As the P_4O_6 vapors are filling the propellant entry chamber cylinders, a polarization field (12), is created by negatively charged electrode (7) and positively charged electrodes (10a, 10b, 10c) which positively polarizes the nucleus of the P_4O_6 vapors within the ion entry chamber cylinders. As this polarization takes place through the P_4O_6 vapors, it creates a separation of a net positive charge of positive ions from a net negative charge of the free electrons within the P_4O_6 vapors. The net charge of positive ions are forced upward toward the negatively charged electrode, and the net charge of the electrons are forced downward toward the positively charged electrode within the P_4O_6 vapors as opposite charges attract and same charges repel.

Also, during the energizing of the polarization field upon the P_4O_6 vapors, there are huge numbers of free mobile electrons (13) within the vapors that pulled away from the P_4O_6 total mass, and are expelled through the bottom of the ion discharge chamber. Due to the removal of these loosely connected electrons by the electrical force of polarization field, this creates an overall a high state of positive ionization of the P_4O_6 vapors.

To further discuss this process, the net positive charge within the P_4O_6 vapors, composed of the nuclei of the P_4O_6 vapors, are attracted upward toward the energized negatively charged polarization electrode (7), and are compressed against the de-energized power electrode (9) which is immediately in front of the negatively charged polarization electrode (7). The net negative charge of the P_4O_6 vapors, composed of the mobile free electrons, are attracted downward toward the energized positively charged polarization electrodes (10a,10b, 10c). Therefore, the function of the polarization field is to create a separation between the net positive charge of the nuclei of the P_4O_6 vapors, and the net negative charge of the mobile free electrons within the P_4O_6 vapors, while inside each of the cylinders (8) of the ion entry chamber cylinders and simultaneously force the positive charged nuclei of the vapor as close as physically possible to the unenergized power electrode (9). This reduces the R in the equation to as close to zero as practical.

An upward force (thrust) is generated by the engine when the polarization field (12) is de-energized (turned off), and immediately the positively charged power electrode (9) is energized while the net positive charge of the P_4O_6 vapors pressed very close to its surface. The P_4O_6 vapors are repelled away from the surface of the energized power electrode (9) with a tremendous velocity, and a huge reaction force is generated perpendicular to the surface of the power electrode (9). The direction opposite of the movement of the P_4O_6 nuclei.

The negative electrodes (11a,11b, 11c) of the power field (14) are located at the bottom of the ion discharge chamber which make up the negative pole of the power field (14).

Pipe 15 connects to manifold 5 to supply the vaporized material. The vaporized material 1 flows through pipe 15 into manifold 5 and through the small orifices 6 into the propellant entry chambers 8. Battery 16 supplies power to valves 2, 3, and 4. Battery 17 supplies power to the electrodes 7, 9, 10, and 11.

When the P_4O_6 vapors are discharged from the bottom of the ion discharge chamber, they recover most of the electrons

and become electrically neutral, and therefore, will not be attracted back to the structure of the craft which would slow the craft down.

See FIG. 4 for diagrammatic process of this action.

Any method of providing high voltage electrical power necessary for the operation of this engine is permitted and acceptable. A time control to alternately energize the polarization field and the power field is necessary to regulate the thrust of the engine over a sustained time.

The directional control of the engine is controlled by the amount of propellant that is injected into each of the three (3) linear sections of the ion discharge chamber. When vapors are repelled out of each section of the ion discharge chamber it creates a reaction Force Vector which is perpendicular to the top surface of that section. The greater the mass of propellant that is repelled from a section, up to a point, the greater the Force Vector generated in that section. There are always three (3) separate Force Vectors acting on the power electrode (9) when the engine is in operation. The engine will always travel in the direction of the Resultant Force composed of the three (3) separate force vectors acting on the engine.

The invention claimed is:

1. An ion engine comprising:

a housing;

an ion discharge chamber divided into a plurality of sections, each section disposed at an angle from the other sections and each section open at a bottom end sufficient to allow the escape of a vaporized material and adapted to connect to a plurality of propellant entry chambers;

a plurality of propellant entry chambers in the housing each open at the end connected to the ion discharge chamber and adapted to receive entry of the vaporized material;

a first positively charged electrode attached to the housing near the bottom end of the ion discharge chamber;

a second positively charged electrode attached to the housing near the bottom end of the ion discharge chamber;

a first negatively charged electrode attached to the housing between the first positively charged electrode and the bottom end of the ion discharge chamber;

a second negatively charged electrode attached to the housing disposed relative to the second positively charged electrode such that the second negatively charged electrode is between the second negatively charged electrode and the bottom end of the ion discharge chamber;

a third positively charged electrode attached to the housing near the area where the propellant entry chambers are adapted to receive the vaporized material;

a third negatively charged electrode attached to the housing near the area where the propellant entry chambers are adapted to receive the vaporized material disposed with the third positively charged electrode between the third negatively charged electrode and each propellant entry chamber;

a means to energize the third negatively charged electrode; and

the vaporized material consisting of P_4O_6.

2. An ion engine comprising;

a housing;

an ion discharge chamber, divided into three sections, each section angled on a slope relative to three dimensional space in the housing open at a bottom end sufficient to allow the escape of a vaporized material and adapted to connect to a plurality of propellant entry chambers;

a plurality of propellant entry chambers in the housing each open at the end connected to the ion discharge chamber and adapted to receive entry of the vaporized material;

7

a first positively charged electrode attached to the housing near the bottom end of the ion discharge chamber;

a second positively charged electrode attached to the housing near the bottom end of the ion discharge chamber;

a means to energize the first and second positively charged electrodes;

a first negatively charged electrode attached to the housing between the first positively charged electrode and the bottom end of the ion discharge chamber;

a second negatively charged electrode attached to the housing disposed relative to the second positively charged electrode such that the second negatively charged electrode is between the second negatively charged electrode and the bottom end of the ion discharge chamber;

a means to energize the first and second negatively charged electrodes;

a third positively charged electrode attached to the housing near the area where the propellant entry chambers are adapted to receive vaporized material;

a third negatively charged electrode disposed close to the second positively charged electrode and with the second positively charged electrode between the third negatively charged electrode and each propellant entry chamber;

a means to energize the third positively charged electrode;

a means to energize the third negatively charged electrode;

a means for supplying vaporized material into the ion entry chamber;

a propellant vaporized material composed of P₄O₆;

a means to regulate the flow of propellant into the propellant entry chambers; and

a means to time the energization and de-energization of the electrodes such that the first and second positive electrodes and the third negative electrode are energized while the first and second negative electrodes and the third positive electrode are not energized and at specified times these electrodes are de-energized and the first and second negative electrodes and the third positive electrode are energized.

3. An ion engine comprising:

a housing;

an ion discharge chamber, divided into three sections, each section angled on a slope relative to three dimensional space in the housing open at a bottom end sufficient to allow the escape of a vaporized material and adapted to connect to a plurality of propellant entry chambers;

a plurality of propellant entry chambers in the housing each open at the end connected to the ion discharge chamber and adapted to receive entry of a vaporized material;

a first positively charged electrode attached to the housing near the bottom end of the ion discharge chamber;

a second positively charged electrode attached to the housing near the bottom end of the ion discharge chamber;

a means to energize the first and second positively charged electrodes;

a first negatively charged electrode attached to the housing between the first positively charged electrode and the bottom end of the ion discharge chamber;

a second negatively charged electrode attached to the housing disposed relative to the second positively charged electrode such that the second negatively discharged electrode is between the second negatively charged and the bottom end of the ion discharge chamber;

8

a battery to energize the first and second negatively charged electrodes;

a third negatively charged electrode disposed close to the third positively charged electrode and with the third positively charged electrode between the third negatively charged electrode and each propellant entry chamber;

a battery to energize the third positively charged electrode;

a battery to energize the third negatively charged electrode;

a pipe adapted to supply the vaporized material into the ion entry chamber;

a vaporized material composed of P₄O₆; and

a valve connected to the pipe and adapted to regulate the flow of vaporized material into the propellant entry chambers.

4. An ion engine comprising:

a housing;

an ion discharge chamber, divided into three sections, each section angled on a slope relative to three dimensional space in the housing, open at a bottom end sufficient to allow the escape of a vaporized material and adapted to connect to a plurality of propellant entry chambers;

a plurality of one inch diameter and one and one half inch deep propellant entry chambers, each open at the end connected to the ion discharge chamber adapted to receive entry of the vaporized material with small orifices in the propellant entry chambers;

a first positively charged electrode attached to the housing near the bottom end of the ion discharge chamber;

a second positively charged electrode attached to the housing near the bottom end of the ion discharge chamber;

a battery to energize the first and second positively charged electrodes;

a first negatively charged electrode attached to the housing between the first positively charged electrode and the bottom end of the ion discharge chamber;

a second negatively charged electrode attached to the housing and disposed relative to the second positively charged electrode such that the second negatively charged electrode is between the second negatively charged and the bottom end of the ion discharge chamber;

a battery to energize the first and second negatively charged electrodes;

a third positively charged electrode attached to the housing near the area where the propellant entry chambers are adapted to receive the vaporized propellant;

a third negatively charged electrode disposed with the second positively charged electrode between the third negatively charged electrode and each propellant entry chamber;

a battery to energize the third positively charged electrode;

a battery to energize the third negatively charged electrode;

a pipe adapted to supply the vaporized material through the pipe;

a manifold attached to the housing connected to the pipe adapted to supply the vaporized material through the manifold and through the small orifices into the ion entry chamber; and

the vaporized material composed of P₄O₆.

* * * * *

HOW TO BE CONTACTED

This has been a very interesting subject to write a book. Had it not been for the Patent No. 7,634,903 B2, *"Ion Impulse Engine"*, it would not have been possible for me to have discovered the reason that the pyramids were built. If you have any questions about any of the design given forth in this book, I can be contacted at:

Richard C. Phillips
Project Alpha-Omega Corporation
P.O.Box 94584
Birmingham, AL
35220

www.ingramcontent.com/pod-product-compliance
Lightning Source LLC
Chambersburg PA
CBHW080653180526
45168CB00008B/3407